PERCENTAGES

Merle Wood
Formerly of Oakland Public Schools
Lafayette, California

Jeanette Powell
Mt. Diablo Unified School District
Concord, California

South-Western Publishing Co.

ISBN: 0-538-70767-4

1 2 3 4 5 6 7 8 9 0 H 98 97 96 95 94 93 92

Printed in the United States of America

Developmental Editor: Penny Shank
Production Editor: Martha G. Conway
Associate Director/Design: Darren Wright
Production Artist: Steve McMahon
Associate Photo Editor/Stylist: Mike O'Donnell
Marketing Manager: Shelly Battenfield

PHOTO CREDITS

Page 3: © Jose Carilla, Ventura, CA

Percentages presents a complete introduction to working problems with percents. Students have many opportunities to practice on the extensive drill exercises. This text-workbook is written specifically for adults and is designed to permit self-paced, individualized instruction and to foster student success.

Percentages progresses from visual identification of percentage parts to working percent word problems in a step-by-step approach. The students are taught to identify the percent of a shape, then write percent values as fractions and decimals. The students learn to figure the percent of a number, find what percent one number is of another, find the number when the percent is known, and solve percent word problems.

SPECIAL FEATURES

Percentages is designed specifically to help you invest in the future of your adult learners and to meet your instructional needs. Some features of the text-workbook include the following:

- A larger typeface is used to make the text-workbook easier for students to use and to read. Pages are colorful and uncrowded.
- Competency-based methodology is used. Clear objectives are presented first, followed by short segments of instruction. These are followed by student activities for immediate reinforcement.
- Content and examples relate to adult-level, real-life issues and skills.
- Pretest and posttests, with answers, are included for self-evaluation.
- Study breaks are included to provide refreshing and useful information that contributes to the general literacy of the student.
- Abundant activities are included, each designed so that students experience frequent and meaningful success.
- Goals are listed for each application to provide motivation and direction.
- All activities are supported with Bonus Activities for students who need a second chance to succeed.
- Answers to all activities are included to facilitate independent, self-paced learning.

- Personal progress is recorded by the student after completing each activity.
- Individual success is measured by evaluation guides in the student's Personal Progress Record.

INSTRUCTOR'S MANUAL ▬▬▬▬▬

The Instructor's Manual provides instructional strategies and specific teaching suggestions for *Percentages,* along with supplementary bonus exercises and answers, additional testing materials, and a certificate of completion.

Bonus Exercises

A bonus exercise matching each activity in the text-workbook is provided in the the Instructor's Manual. These bonus exercises make it possible for students to have a second chance to reach the goals set for each activity. Answers to the bonus exercises are provided in the manual. These materials may be reproduced for classroom use.

Testing Materials

Two additional tests are provided in the text-workbook to allow for more flexible instruction and evaluation.

Certificate of Completion

Upon completion of *Percentages,* a student's success may be recognized through a certificate of completion. This certificate lists the skills and topics covered in this text-workbook. A certificate master is included in the manual.

▲▲▲ CONTENTS

GETTING ACQUAINTED

Even though many of us do not use percents every day, percents are needed every once in a while either at work or at home. We need to know how to solve problems with percents.

You may be enrolled in a class using *Percentages* because you know your math skill is in need of improvement. Or you might be taking this class just to brush up on your math skills. This book will help you to reach either objective.

HOW YOU WILL LEARN

Percentages is written with you in mind. You will learn just what percentages are. You will also learn to add, subtract, multiply, and divide percentages. You will work many practice problems to help you understand exactly how to work with percentage problems.

Learn at Your Own Pace

You will progress through this text-workbook at your own pace. You may move ahead faster, or go slower, than other students. But don't be concerned about this. You are to work at *your* best speed.

Learn Skills Successfully

You are given objectives and goals for each unit. You will know what you are to accomplish. You will study a topic. Then you will complete an exercise. This lets you drill over what you have just learned. When you have shown that you know the topic, you will move on to the next topic. You will always know just how well you are doing as you move through each step in this book.

Complete Bonus Exercises

You may not reach your assigned goal on every practice exercise. When this happens, you should review the lesson and then do a Bonus Exercise. These exercises cover the same lessons as the practice exercises in the book. They give you a second chance to reach your goal. When you score higher on a Bonus Exercise than you did on the original activity you may change your score on your Personal Progress Record. Your instructor has copies of the Bonus Exercises and the answers to them.

Check Your Own Success

You will keep track of your own success. You will check all of your own work. The answers are near the back of the book. The color pages make them easy to find. Always do the exercises *before* you look at the answers. Use the answers as a tool to verify your work—not as a means of filling in the blanks. You will record your scores on your own Personal Progress Record, which is also at the back of the book.

WHAT YOU WILL LEARN ━━━━━━━━

As you study *Percentages,* you will gain an understanding of percent values and will learn how to solve problems that have percentages in them.

In Unit 1, Understanding Percents, you will learn to identify the percent of a shape, state the percent of money, and write percent value as a decimal or fraction.

In Unit 2, Finding the Percent of a Number, you will learn to find the percent of a number, estimate percents, and solve percent word problems.

In Unit 3, Finding Other Kinds of Percents, you will learn to find the percent of a number, find what percent one number is of another, find a number when the percent of the number is known, and figure percent word problems.

SPECIAL FEATURES ━━━━━━━━

Percentages has a number of special features. These features will help you learn and then use the material successfully.

Checking What You Know

You can check the percent skill you already have before you begin your study in this book. *Checking What You Know* lets you determine what percent skills you have and those you need to improve upon. Then, when you complete the book, you will do an exercise called *Checking What You Learned.* By comparing these two scores, you will see how much you have gained through your study.

Learning Number Facts

At various points throughout this text-workbook you will find brief items titled *It's a Fact.* These inserts contain tables or other numeric or general information that will be of interest to you. They provide a change of pace from your regular learning.

Putting It Together

Each unit has a number of short exercises called Checkpoints. These exercises will help you check your understanding of a specific topic before going on. At the end of each unit you will find a section titled *Putting It Together*. This section contains several exercises. They are similar to the Checkpoints within the units. They will help you to reinforce the skills you learned in each unit.

Personal Progress Record

You will keep track of your own progress. Once you check your answers, you will record your score on your *Personal Progress Record* at the end of this book. After you finish a unit, you will be able to see your level of success.

Completion Certificate

When you finish your study in this book, you may be eligible for a certificate of completion. Your instructor will explain the skill level required for this award.

READY TO START!

You are now ready to start improving your ability to solve problems dealing with percentages. Through your study and completion of the exercises, you should quickly develop an improved ability to do percent problems.

Your new skill will prove to be of benefit to you. You can do both your personal and on-the-job computing with added confidence. You may also have improved opportunities to move up the job ladder where, in many cases, well-developed math skills are required.

Turn to page xi and complete Checking What You Know. Check your answers with the answers on page 41. Then begin Unit 1, Understanding Percents.

CHECKING WHAT YOU KNOW

Take this pretest before starting *Percentages*. The 25 questions will tell you how much you already know about percents. They will also give you an idea of those skills that you need to improve. There is no time limit, so take your time.

Follow the directions carefully. Write your answers in the spaces provided. When you finish, check your work with the answers near the back of this book. Give yourself 4 points for each correct answer. Then record your score in your Personal Progress Record. Use the analysis chart to see where you need to improve your skill.

DIRECTIONS: Change the following amounts to decimals, fractions, and percents.

	Decimal	Fraction	Percent	● ___.07___
Seven hundredths =	● 0.07	**(1)**	**(2)**	1. _____
				2. _____
Thirty hundredths =	**(3)**	**(4)**	**(5)**	3. _____
				4. _____
One tenth =	**(6)**	**(7)**	**(8)**	5. _____
				6. _____
Nine thousandths	**(9)**	**(10)**	**(11)**	7. _____
				8. _____

Change to percents.

9. _____

(12) $\frac{3}{8}$ **(13)** $\frac{3}{5}$ **(14)** $\frac{1}{2}$ **(15)** $\frac{5}{8}$

10. _____

11. _____

12. _____

13. _____

14. _____

15. _____

Work the following problems.

(16) 30% of 150   **(17)** 3% of 80   **(18)** 18% of 45

(19) What percent of 80 is 20?

(20) What percent of 125 is 25?

(21) 12 is 20% of what number?

(22) 35 is 25% of what number?

(23) How much would a 15% tip be on an $8.00 dinner?

(24) Marchetta's rent increased 5%. She was paying $750 per month. How much will she have to pay after the rent increase?

(25) Mr. Schumacher won 75% of the games he played. If he won 15 games, what was the total number of games he played?

16. _____

17. _____

18. _____

19. _____

20. _____

21. _____

22. _____

23. _____

24. _____

25. _____

☞ *Check your work on page 41. Record your score on page 45.*

UNIT 1

Understanding Percents

WHAT YOU WILL LEARN

When you finish this unit, you will be able to:
- Identify the percent of a shape.
- State the percent of a given amount of money.
- Write a percent value as a decimal or fraction.

Statements such as "He's batting fifty percent today," or "There is a 60 percent chance of rain" give information regarding conditions of interest to us. In this unit you will study the meaning of percent to better understand it in your daily life.

IDENTIFYING PART AND WHOLE

Percent means something has been divided into 100 parts. The symbol used for percent is **%.** Twenty percent (20%) means 20 out of 100. An easy way to think of a percent is as a fraction. Since percent means 100, use 100 as the denominator of a fraction. The part is 20, so it is the numerator of the fraction.

In Illustration 1-1, square A shows 20 of the 100 squares shaded, so 20% is shaded. If you know the percent of part of an object, the percent of the rest of the object may be calculated by subtracting the known percent from 100.

A.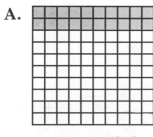

20% shaded

B.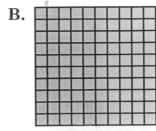

100% shaded

C.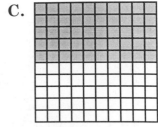

50% shaded

Square B + Square C = 150%

Illustration 1-1

1

One whole object is 100%. Two objects are 200%. If there is one whole object and half of another object, the two together equal 150%. In Illustration 1-1, squares B and C together equal 150%.

Circle graphs are frequently used in newspapers to display statistics. Circle A in Illustration 1-2 is divided into three unequal parts. The three parts total 100%. To find the missing percent of a circle, add the known percents and subtract from 100, as shown in circle B in Illustration 1-2.

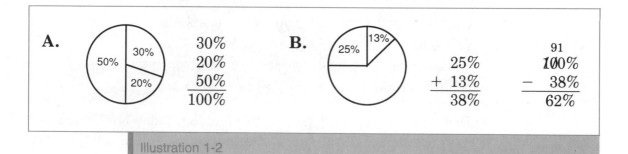

Illustration 1-2

✓ CHECKPOINT 1-1

YOUR GOAL:
Get 6 or more answers correct.

1. Identify the percent of the shaded area of the following squares. The first one has been done as an example.

$$\frac{50}{100} = 50\%$$

A.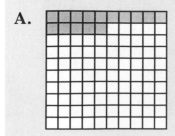

B.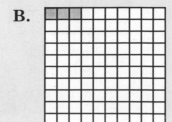

C.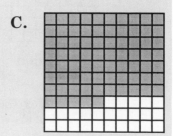

_____   _____   _____

2. Write the missing percent in each of the circle graphs that follow. The first one has been done as an example.

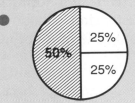

A.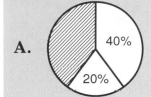

B.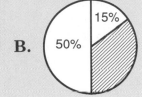

C.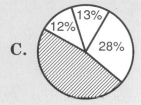

3. Answer the following word problems in the spaces provided.

A. A farmer with a circular sprinkler system planted 25% of her garden in squash, 15% in beans, 10% in carrots, and 30% in tomatoes. What percent of the garden was left for potatoes?

_____

B. The Scott family developed a budget which allowed 40% for housing and utilities, 30% for food, 5% for transportation, 10% for clothing, and 5% for miscellaneous. What percent was available to save for a vacation?

_____

☞ *Check your work on page 41. Record your score on page 46.*

PERCENT AND MONEY

A dollar equals 100 pennies. Figuring the percent of a dollar is like figuring the percent of 100 squares. Illustration 1-3 shows 100 pennies equal 100% and 20 pennies equal 20% of a dollar. Two dollars is 200%.

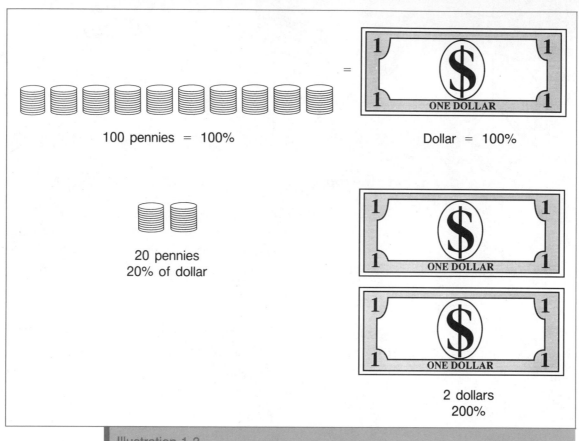

100 pennies = 100%   Dollar = 100%

20 pennies
20% of dollar

2 dollars
200%

Illustration 1-3

✔ CHECKPOINT 1-2

YOUR GOAL:
Get 5 or more answers correct.

Write the percent of a dollar of the following amounts of money. The first one has been done as an example.

 $3.00

    300%

1. Twenty cents

2. $1.50

3. Seventy-five cents

4. Three cents

_____   _____   _____   _____

5. Two dollars	6. $0.84	7. Thirty cents
_____	_____	_____

 Check your work on page 41. Record your score on page 46.

EQUIVALENTS: DECIMALS, FRACTIONS, AND PERCENTS

A value of less than one may be written as a fraction, decimal, or percent. The following numbers are all equivalent values. **Equivalent** values are all equal.

$$.25 = \frac{25}{100} = 25\%$$

In order to calculate percents it is necessary to be able to convert decimals, fractions, and percents to their equivalent values.

Change from Decimal to Percent

To change a decimal to a percent, move the decimal point two places to the right. Write the percent sign to the right of the last digit. Illustration 1-4 shows that you must always move the decimal two places to the *right* and that the zero may be dropped in front of a whole number.

Illustration 1-4

0.35 = 35%	0.137 = 13.7%
0.09 = 9%	0.4782 = 47.82%

✔ CHECKPOINT 1-3

YOUR GOAL:
Get 6 or more answers correct.

Change the following decimals to percents. The first one has been done as an example.

● 0.08 = __8%__

1. 0.12 = _____
2. 2.35 = _____
3. 3.5 = _____

4. 0.38 = _____
5. 0.07 = _____
6. 1.00 = _____
7. 0.90 = _____

 Check your work on page 41. Record your score on page 46.

Change from Percent to Decimal ▬▬▬▬

To change from a percent to a decimal, move the decimal point two places to the *left*. Drop the percent sign. For one-digit percents, write a zero between the decimal point and the digit, as shown in the 5% example in Illustration 1-5.

Illustration 1-5

5% = .05 = 0.05	12% = .12 = 0.12
30% = .30 = 0.30	125% = 1.25 = 1.25

CHECKPOINT 1-4

YOUR GOAL:
Get 4 or more answers correct.

Change the following percents to decimals. The first one has been done as an example.

●   37% = __0.37__

1.   21% = _____

2. 250% = _____

3.   3% = _____

4.   40% = _____

5. 115% = _____

 Check your work on page 41. Record your score on page 46.

◄ IT'S A FACT—Tests Required

The process of getting many jobs in today's world involves passing reading, writing, and math tests. Apprentice programs for many jobs require good math skills.

Change from Decimal to Fraction ▬▬▬▬

To change a decimal to a fraction, simply read the decimal number. Illustration 1-6 shows some decimal numbers in words and fractions.

Illustration 1-6

$$.5 = \text{five tenths} = \frac{5}{10}$$

$$.32 = \text{thirty-two hundredths} = \frac{32}{100}$$

$$.006 = \text{six thousandths} = \frac{6}{1000}$$

 CHECKPOINT 1-5

YOUR GOAL:
Get 13 or more answers correct.

Complete the equivalent values chart that follows. The first one has been done as an example.

	Decimals	Words	Fractions
●	0.18	**eighteen hundredths**	$\frac{18}{100}$
1.	0.5		
2.	0.030		
3.	0.25		
4.	0.99		
5.	0.7		
6.	0.128		
7.	0.09		
8.	0.1		

☞ *Check your work on page 41. Record your score on page 46.*

Changing Fractions to Decimals and Percents

The fractions $\frac{1}{4}$, $\frac{1}{2}$, and $\frac{3}{4}$ are easy to convert to equivalent fractions and then percents, or from percents to equivalent fractions. To convert from a fraction to a percent, find a number that when multiplied by the denominator will equal 100. Then multiply both the numerator and the denominator by that number, as shown in Illustration 1-7.

Illustration 1-7

$$\frac{1}{4} \binom{\times\ 25}{\times\ 25} = \frac{25}{100} = 25\% \qquad\qquad \frac{1}{2} \binom{\times\ 50}{\times\ 50} = \frac{50}{100} = 50\%$$

$$\frac{3}{4} \binom{\times\ 25}{\times\ 25} = \frac{75}{100} = 75\%$$

Changing Percents to Fractions

To change from a percent to an equivalent fraction, write the percent as a fraction. Then reduce the fraction to its lowest term, as shown in Illustration 1-8. Notice that you divide when reducing fractions.

Illustration 1-8

$$25\% = \frac{25}{100} \binom{\div\ 25}{\div\ 25} = \frac{1}{4} \qquad\qquad 50\% = \frac{50}{100} \binom{\div\ 50}{\div\ 50} = \frac{1}{2}$$

CHECKPOINT 1-6

YOUR GOAL:
Get 4 or more answers correct.

1. Convert the following fractions to equivalent fractions with 100 as the denominator. Then change to percents. The first one has been done as an example.

   ● $\frac{3}{10} \binom{\times\ 10}{\times\ 10} = \frac{30}{100} =$ __30%__

   A. $\frac{4}{20}$ = _____

   B. $\frac{2}{5}$ = _____

   C. $\frac{8}{50}$ = _____

2. Convert the following percents to equivalent fractions with 100 as the denominator. Reduce to its lowest term. The first one has been done as an example.

   ● $75\% = \frac{75}{100} \binom{\div\ 25}{\div\ 25} = \frac{3}{4}$

   A. $50\% = $ —— $\binom{\div}{\div}\ \ \} = $ ——

   B. $60\% = $ —— $\binom{\div}{\div}\ \ \} = $ ——

   C. $20\% = $ —— $\binom{\div}{\div}\ \ \} = $ ——

☞ *Check your work on page 41. Record your score on page 46.*

Changing Fractions to Percents by Dividing

Another way to change a fraction to a percent is to divide the denominator into the numerator, adding a decimal and two zeros after the numerator. The answer is a decimal number. To change the decimal number to a percent, move the decimal two places to the *right* and add a percent sign as shown in Illustration 1-9.

Illustration 1-9

$$\frac{1}{4} = 4\overline{)1.00} \quad \begin{array}{c} .25 \\ \end{array} = .25 = 25\%$$

A calculator may be used to convert a fraction to a decimal. Simply divide the numerator by the denominator as shown in Illustration 1-10.

Illustration 1-10

To change $\frac{1}{5}$ to a decimal

Key: **1** ÷ **5** = 0.2

The answer, 0.2, should appear in the display. When converted to a percent, 0.2 = 20%.

IT'S A FACT—Management's Concern Is Employees

Today many companies are taking a greater interest in their employees' lives. Some companies provide on-site child care, some companies provide counseling services, and other companies provide educational opportunities. Human resources is an occupational field that is expanding.

✔ CHECKPOINT 1-7

YOUR GOAL:
Get 2 or more answers correct.

Change the following fractions to percents by dividing. Round the decimal to the nearest thousandth. The first one has been done as an example.

● $\frac{5}{8}$ ___0.625___ $8\overline{)5.000}$ (0.625)

1. $\frac{2}{5}$ _____

2. $\frac{3}{7}$ _____

3. $\frac{5}{6}$ _____

☞ *Check your work on page 41. Record your score on page 46.*

Change Percents with Fractions to Decimals

To change a percent with a fraction to a decimal requires two steps. First the fraction must be changed to a decimal number. Then the decimal can be moved two places to the left. In Illustration 1-11, the fraction $\frac{1}{4}$ was changed to the decimal 0.25. Then the decimal was moved two places to the left. Notice that a zero had to be placed before the 7 so that the decimal could move two places to the left.

Illustration 1-11

$$7\frac{1}{4}\% = 7.25\% = 0.0725$$

CHECKPOINT 1-8

Change the following percents with fractions to decimals. The first one has been done as an example.

● $6\frac{1}{2}\%$ =   6.5%   =   **0.065**

1. $12\frac{1}{8}\%$ = _____

2. $6\frac{2}{5}\%$ = _____

3. $6\frac{1}{2}\%$ = _____

4. $75\frac{3}{4}\%$ = _____

5. $20\frac{3}{8}\%$ = _____

 Check your work on page 41. Record your score on page 46.

◀ IT'S A FACT—Areas of Opportunity

Occupational opportunities in technical areas are growing rapidly; however, they will account for only about 4% of the new jobs. Clerical and service occupations will have about a 40% employment growth.

WHAT YOU HAVE LEARNED

In this unit you learned how to identify the percent of parts of a shape. The activities guided you through the concept that a whole is 100%. More than a whole is indicated by a number such as 200%. A dollar is easy to relate to a percent. It takes 100 pennies (part) to equal one dollar (whole) or 100%. Percent values can be written as a decimal or fraction.

ACTIVITY 1-1   YOUR GOAL: Get 6 or more answers correct.

1. Write a fraction indicating the number of shaded squares over the total squares. Then change the fraction to a percent. The first one has been done as an example.

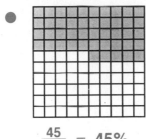

$$\frac{45}{100} = 45\%$$

A.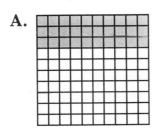

B.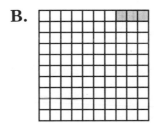

C.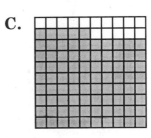

_____   _____   _____

2. Write the missing percent in each of the following circle graphs. The first one has been done as an example.

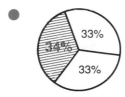

A.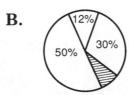

B.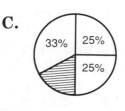

C.

3. A family was sewing curtains for their motor home. They needed 37% of the fabric for the windshield, 30% for the right-side window, and 25% for the left-side window. What percent of the fabric was left for the rear window?

_____

☞ *Check your work on page 41. Record your score on page 46.*

11

ACTIVITY 1-2   YOUR GOAL: Get 2 or more answers correct.

Write the percent of a dollar for the following amounts of money. The first one has been done as an example.

● $5.00

   <u>   500%   </u>

1. $0.65   **2.** Thirty cents   **3.** Three cents

   _____   _____   _____

☞ *Check your work on page 42. Record your score on page 46.*

ACTIVITY 1-3   YOUR GOAL: Get all 3 answers correct.

Change the following decimals to percents. The first one has been done as an example.

● 0.14 = <u>   14%   </u>

1. 0.09 = _____

2. 2. = _____

3. 3.7 = _____

☞ *Check your work on page 42. Record your score on page 46.*

ACTIVITY 1-4   YOUR GOAL: Get all 3 answers correct.

Change the following percents to decimals. The first one has been done as an example.

● 30% = <u>   0.30   </u>

1. 128% = _____

2. 2% = _____

3. 27% = _____

☞ *Check your work on page 42. Record your score on page 46.*

ACTIVITY 1-5   YOUR GOAL: Get 6 or more answers correct.

Complete the equivalent values chart that follows. The first one has been done as an example.

	Decimals	Words	Fractions
●	0.19	nineteen hundredths	$\frac{19}{100}$
1.	0.7		
2.	0.27		
3.	0.40		
4.	0.05		

☞ *Check your work on page 42. Record your score on page 46.*

ACTIVITY 1-6   YOUR GOAL: Get 5 or more answers correct.

1. Convert the following fractions to equivalent fractions with 100 as the denominator. Then change to percents. The first one has been done as an example.

   ● $\frac{8}{10}$ = $\frac{80}{100}$ = ___80%___

   A. $\frac{3}{5}$ = _____

   B. $\frac{4}{20}$ = _____

   C. $\frac{7}{20}$ = _____

2. Convert the following percents to equivalent fractions with the numerator over 100. Reduce to lowest terms when possible. The first one has been done as an example.

   ● 25% = $\frac{25}{100}$ = ___$\frac{1}{4}$___

   A. 80% = _____

   B. 5% = _____

   C. 13% = _____

☞ *Check your work on page 42. Record your score on page 46.*

ACTIVITY 1-7   YOUR GOAL: Get 2 or more answers correct.

Change the following fractions to percents. Round to the nearest tenth.
The first one has been done as an example.

● $\frac{2}{7} = 7\overline{)2.00}^{.28} = .30 \quad = \underline{\ 30\%\ }$

1. $\frac{3}{8}$ _____

2. $\frac{1}{6}$ _____

3. $\frac{1}{3}$ _____

ACTIVITY 1-8   YOUR GOAL: Get 3 or more answers correct.

Change the following percents with fractions to decimals. Round to the nearest
thousandth. The first one has been done as an example.

● $15\frac{1}{8}\% = \underline{\ 0.151\ }$

1. $3\frac{1}{2}\% = \underline{\ \ \ \ \ }$

2. $9\frac{3}{5}\% = \underline{\ \ \ \ \ }$

3. $40\frac{3}{4}\% = \underline{\ \ \ \ \ }$

4. $21\frac{2}{7}\% = \underline{\ \ \ \ \ }$

5. $201\frac{1}{3}\% = \underline{\ \ \ \ \ }$

☞ *Check your work on page 42. Record your score on page 46.*

UNIT 2

Finding the Percent of a Number

WHAT YOU WILL LEARN

When you finish this unit, you will be able to:
- Find the percent of a number.
- Calculate percents with mixed numbers.
- Estimate percents.
- Solve percents in word problems.

Percents are important because they are a factor in our everyday lives. Each time we buy something we pay sales tax. When we put gas in our car or truck we pay a gas tax. All these taxes are figured by multiplying the amount of sale by a certain percent.

IT'S A FACT—Where You'll Find Jobs

The U.S. Department of Labor predicts that the following five areas of employment will have the largest percentage of new job gains during the next ten years: Paralegals, Medical Assistants, Physical Therapists, Data Processing Equipment Repairers, and Home Health Aids.

FIND THE PERCENT OF A NUMBER

The most common situation in which percents are used is finding the percent of a number. To find the percent of a number, change the percent to a decimal and multiply. When changing from a percent to a decimal, remember the decimal must move two places to the left. If the percent part of the problem does not have a decimal, a decimal may be placed to the right of the last number. If the percent is a one-digit number, a zero must be placed to the left of the number before you can move the decimal two places. Study the examples in Illustration 2-1.

Illustration 2-1

Find 40% of 275	Find 8% of 3.50
40% = 0.40	8% = 0.08
0.40 × 275 = 110	0.08 × 3.50 = 0.28

This method of calculating percent is extremely easy to do on the calculator. Simply key the problem as shown here.

What is 15% of 35?   Key **35 × 15%**

The answer will appear in the display window after pressing the equal key. If your calculator doesn't have a percent key, simply change the percent to a decimal and multiply.

CHECKPOINT 2-1

YOUR GOAL:
Get 5 or more answers correct.

Find the percent of the following numbers *without* using a calculator. Complete your work on a separate piece of paper. Write your answers in the spaces provided. The first one has been done as an example.

● 6% of 130 = 0.06 × 130 =   <u>7.8</u>

1. 18% of $0.50 =   _____

2. 5% of 140 =   _____

3. 30% of $250.00 =   _____

4. 20% of 150 =   _____

5. 8% of $350 =   _____

6. 25% of 45 =   _____

 Check your work on page 42. Record your score on page 46.

CALCULATING PERCENTS WITH MIXED NUMBERS

Sales tax amounts are frequently figured as mixed numbers such as $7\frac{3}{4}\%$. In those cases, the fraction must be changed to a decimal number before the decimal point is moved two places to the left. In Illustration 2-2, the fraction $\frac{3}{4}$ is changed to 0.75 before the decimal point is moved two places to the left. Notice a zero was placed in front of the 7 before the decimal point was moved.

Illustration 2-2

$$7\frac{3}{4}\%$$
$$\downarrow$$
$$7.75\% = 0.0775$$

If you cannot change the fraction to a decimal in your mind, then divide the numerator by the denominator. You will have to add a decimal and two zeros to the numerator before you can divide. Fraction percents that do not divide evenly are usually rounded to the nearest hundredth. Study Illustration 2-3.

Illustration 2-3

$$4\frac{1}{5}\%$$
$$\downarrow$$

$$\begin{array}{r} .20 \\ 5{\overline{\smash{)}\,1.00}} \\ \underline{1\,0} \\ 0 \end{array} = 4.20\%$$

$$\frac{1}{3}\%$$
$$\downarrow$$

$$\begin{array}{r} .333 \\ 3{\overline{\smash{)}\,1.000}} \\ \underline{-9} \\ 10 \\ \underline{-9} \\ 10 \\ \underline{-9} \\ 1 \end{array} = .33\%$$

✔ CHECKPOINT 2-2

YOUR GOAL:
Get 5 or more answers correct.

On a separate piece of paper, calculate the percent for the following fractions and mixed numbers. Round the answer to the nearest tenth. Write your answers in the spaces provided. The first one has been done as an example.

● $8\frac{4}{5}\%$ of 20 = 8.8% of 20 = 0.088 × 20 = 1.760 =   __1.8__

1. $12\frac{1}{4}\%$ of 350 =   _____

2. $18\frac{1}{2}\%$ of 2000 =   _____

3. $\frac{1}{3}\%$ of 6 =   _____

4. $5\frac{2}{3}\%$ of 20 =   _____

5. $4\frac{1}{2}\%$ of 100 =   _____

6. $7\frac{3}{4}\%$ of 75 =   _____

☞ **Check your work on page 42. Record your score on page 46.**

◀ **IT'S A FACT—Increase in Health Care Jobs**

In certain parts of the country the health care industry is projected to grow by about 23%. The following are estimates of percent of job increases: Registered Nurses (30%), LVN (27%), Nurse Assistant (16%), and Home Health Aide (60%).

ESTIMATING PERCENTS

Percents such as 50% and 25% are easy to estimate. You know that 50% is $\frac{1}{2}$. Therefore, to estimate 50% of a number divide the number by 2.

$$50\% \text{ of } 300 = 150 \qquad 2\overline{)300}^{\,150}$$

You know that 25% is $\frac{1}{4}$. Therefore, to estimate 25% of a number, divide the number by 4.

$$25\% \text{ of } 16.32 = 4.08 \qquad 4\overline{)16.32}^{\,4.08}$$

✔ **CHECKPOINT 2-3**

YOUR GOAL:
Get 5 or more answers correct.

Calculate the following percents by dividing by 2 or 4. Round to the nearest tenth. Do your dividing on a separate piece of paper. Write your answers in the spaces provided. The first one has been done as an example.

● 25% of 250 = __62.5__ $4\overline{)250.0}^{\,62.5}$

1. 50% of 80 = _____

2. 25% of 200 = _____

3. 25% of 808 = _____

4. 50% of 380 = _____

5. 50% of 35 = _____

6. 25% of 40 = _____

☞ *Check your work on page 42. Record your score on page 46.*

Estimating percents and doing mental math is very convenient. When you are able to estimate a 10 or 15 percent tip at a restaurant, you know how much money to leave. When shopping at a store with a percent-off sale, you can estimate how much will be taken off the sales price. Mental math and estimation are skills that can be learned. When you think of percent, think of something divided into 100 pieces. Study the charts in Illustration 2-4. Look for patterns.

Illustration 2-4

100% is 100 out of 100
90% is 90 out of 100
80% is 80 out of 100

100% is 100 out of 100
10% is 10 out of 100
1% is 1 out of 100

25% of 200 = 50	(200 ÷ 4)
50% of 200 = 100	(200 ÷ 2)
100% of 200 = 200	(200 ÷ 1)
200% of 200 = 400	(200 × 2)

CHECKPOINT 2-4

YOUR GOAL:
Get 7 or more answers correct.

Find the percent of the following numbers. The first one has been done as an example.

● 1% of 400 = ___4___

1. 10% of 400 = _____   5. 10% of 300 = _____

2. 25% of 400 = _____   6. 25% of 300 = _____

3. 50% of 400 = _____   7. 50% of 300 = _____

4. 100% of 400 = _____   8. 1% of 300 = _____

 Check your work on page 42. Record your score on page 46.

FIGURING PERCENTS IN WORD PROBLEMS

The type of percent most often used is the percent of a number. In simple problems such as "What is 50% of $75?" simply do mental math and divide by 2 or figure the percentage by changing the percent to a decimal and multiplying. Some problems require more than one step. In problems requiring more than one step you might ask yourself the following questions:

1. What is being asked?
2. What are the key words?

3. What has to be done first?

4. Should I add or subtract after I find the percent?

Study the word problem in Illustration 2-5.

Illustration 2-5

A sweater that normally sells for $45.00 is on sale for 20% off. What is the price of the sweater?

1. The question is, what will you have to pay for the sweater?
2. The key word is "off."
3. You must find 20% of $45.00 first.
4. Then subtract the answer from $45.00.

20% of $45.00
$0.20 \times \$45.00 = \$9.00$
$\$45.00 - \$9.00 = \$36.00 =$ price of sweater

Illustration 2-6

Stores often sell items at a certain percent off the regular price.

CHECKPOINT 2-5

YOUR GOAL:
Get 5 or more answers correct.

Find the answers to the following word problems. Use a separate piece of paper if necessary. Write your answers in the spaces provided. The first one has been done as an example.

● Jose got a $5\frac{1}{4}\%$ raise. His former rate of pay was $8.25 per hour. What is his new rate of pay?

$$
\begin{array}{r}
\$8.25 \\
\times\ .0525 \\
\hline
4125 \\
1650 \\
4125 \\
\hline
0.433125
\end{array}
$$

$$
\begin{array}{r}
\$8.25 \\
+\ 0.43 \\
\hline
\$8.68
\end{array} = \text{new rate of pay}
$$

$8.68

1. Felix wanted to buy a fishing pole that cost $89.00. The tax was $6\frac{1}{2}\%$. What was the total cost of the fishing pole?

    _____

2. The Manual family went out for a pizza dinner that cost $23.00. They gave the waitress a 15% tip. How much tip did she receive?

    _____

3. Barbara borrowed $5,000 to buy a car. The credit union charged 12% interest. What is the total amount Barbara had to repay the credit union?

    _____

4. A living room set was advertised at $800 with a 40% discount. What was the sale price of the set?

    _____

5. Bud purchased a used car for $1,500. He had to pay $5\frac{1}{2}\%$ sales tax. How much did the car cost altogether?

    _____

6. Jeanette was paid $18.00 for baby-sitting. She was given a 10% tip. How much was her tip?

    _____

☞ *Check your work on page 42. Record your score on page 46.*

WHAT YOU HAVE LEARNED

Now that you have finished this unit you will be able to calculate the percent of a number or key a calculator to find the percent of a number. You also practiced figuring one- and two-step word problems.

ACTIVITY 2-1   YOUR GOAL: Get 5 or more answers correct.

Find the percent of the following numbers *without* using a calculator. Complete your work on a separate piece of paper. Write your answers in the spaces provided. The first one has been done as an example.

-    7% of 150 = 0.07 $\times$ 150 =   __10.5__

1. 8% of 200 =   _____

2. 16% of 35 =   _____

3. 40% of 100 =   _____

4. 35% of 150 =   _____

5. 10% of 35 =   _____

6. 50% of 800 =   _____

☞ *Check your work on page 42. Record your score on page 46.*

ACTIVITY 2-2   YOUR GOAL: Get 5 or more answers correct.

On a separate piece of paper, calculate the percent for the following fractions and mixed numbers. Round the answer to the nearest tenth. Write your answers in the spaces provided. The first one has been done as an example.

-    $7\frac{2}{7}$% of 500 = 7.3% $\times$ 500 =   __36.5__

1. $15\frac{3}{4}$% of 500 =   _____

2. $\frac{1}{2}$% of 50 =   _____

3. $8\frac{1}{5}$% of 25 =   _____

4. $24\frac{1}{4}$% of 80 =   _____

5. $12\frac{1}{2}$% of 82 =   _____

6. $20\frac{1}{4}$% of 36 =   _____

☞ *Check your work on page 42. Record your score on page 46.*

ACTIVITY 2-3   YOUR GOAL: Get 5 or more answers correct.

Calculate the following percents by dividing by 2 or 4. Round to the nearest tenth. Do your dividing on a separate piece of paper. Write your answers in the spaces provided. The first one has been done as an example.

● 50% of $175.00 = __$87.50__

$$\begin{array}{r} 87.50 \\ 2\overline{)175.00} \end{array}$$

1. 25% of 20 = _____

2. 50% of 25 = _____

3. 50% of 300 = _____

4. 25% of 1000 = _____

5. 25% of 80 = _____

6. 50% of 90 = _____

☞ *Check your work on page 43. Record your score on page 46.*

ACTIVITY 2-4   YOUR GOAL: Get 7 or more answers correct.

Calculate the percent of the following numbers. The first one has been done as an example.

● 10% of 300 = __30__

1. 1% of 300 = _____   **5.** 10% of 500 = _____

2. 25% of 800 = _____   **6.** 25% of 500 = _____

3. 50% of 26 = _____   **7.** 50% of 500 = _____

4. 100% of 380 = _____   **8.** 1% of 500 = _____

☞ *Check your work on page 43. Record your score on page 46.*

ACTIVITY 2-5   YOUR GOAL: Get 5 or more answers correct.

Find the answers to the following word problems. Use a separate piece of paper if necessary. Write your answers in the spaces provided. The first one has been done as an example.

● Ms. Martinez is shopping for new shoes. The shoes she likes cost $48.00 and are on sale for 30% off. How much will Ms. Martinez pay for her shoes?

$$
\begin{array}{rr}
\$48. \\
\times\ 0.30 \\
\hline
00 \\
14\ 4 \\
\hline
14.40
\end{array}
\qquad
\begin{array}{r}
\$48.00 \\
-\ 14.40 \\
\hline
\$33.60\ =\ \text{cost of shoes}
\end{array}
\qquad
\underline{\textbf{\$33.60}}
$$

1. Bill's hardware purchase came to $28.00. He must pay a $5\frac{1}{4}\%$ sales tax. How much sales tax must he pay?

   _____

2. Rebecca and Jessica's lunch bill was $12.50. They want to leave a 15% tip. How much money should they leave altogether?

   _____

3. Joan's car loan was $10,000. Her credit union charges 15% interest. How much interest will she have to pay?

   _____

4. The nursery had a 30% off clearance sale on plants. What was the price of a tree that normally cost $15.00?

   _____

5. David took his girlfriend out to dinner. Their dinner cost $62.00. He left a 15% tip. How much did he pay altogether?

   _____

6. Sue and Neva paid $35 for a boat part. The sales tax was $7\frac{1}{2}\%$. What was the total amount of their purchase?

   _____

☞ *Check your work on page 43. Record your score on page 46.*

UNIT 3

Finding Other Kinds of Percents

WHAT YOU WILL LEARN

When you finish this unit, you will be able to:
- Find the percent of a number.
- Find what percent one number is of another.
- Find the number when the percent is known.
- Solve percent word problems.

Being able to figure percents is especially important when you borrow money. The interest charged varies from lender to lender. Being able to calculate and compare the cost of interest can save you money.

Illustration 3-1

Being able to figure percents is especially important when you borrow money.

FINDING THE PERCENT OF A NUMBER

There are two easy methods to figure the percent of a number:
1. Change the percent to a decimal and multiply.
2. Use the percent key on a calculator.

Sometimes, however, the percent is not known and you must find the percent. Sometimes the total number is unknown. The method, *percent over 100 is equal to the part over the total,* can be used with all three kinds of percent problems. In Illustration 3-2, the formula and steps for finding the percent of a number are shown.

Illustration 3-2

What is 30% of 150?

Formula	Step 1:	Step 2:
	Cross multiply	Divide the result of cross multiplying by the remaining number

$$\frac{\%}{100} = \frac{Part}{Total}$$

$$\frac{30}{100} \diagdown \overline{150}$$

$$30 \times 150 = 4,500$$

$$\begin{array}{r} 45 \\ 100\overline{)4500} \\ \underline{400} \\ 500 \\ \underline{500} \end{array}$$

Answer: 45 is 30% of 150.

When a number is divided by 100, the answer is the same as if you moved the decimal point two places to the left. A decimal may be added to the right of the last digit in a whole number without changing the value of the number. Dividing by 100 makes the number smaller. Illustration 3-3 shows a whole number and a decimal number divided by 100.

Illustration 3-3

$$\begin{array}{r} 45 \\ 100\overline{)4500} \\ \underline{400} \\ 500 \\ \underline{500} \end{array}$$ or 4,500 ÷ 100 = 45 00. = 45

$$\begin{array}{r} 0.5218 \\ 100\overline{)52.1800} \\ \underline{50\ 0} \\ 2\ 18 \\ \underline{2\ 00} \\ 180 \\ \underline{100} \\ 800 \\ \underline{800} \end{array}$$ or 52.18 ÷ 100 = 52.18 = 0.5218

If the percent is a mixed number, the fraction must be changed to a decimal before putting it over 100 as shown in Illustration 3-4.

Illustration 3-4

What is $7\frac{1}{4}$% of 6?

$7\frac{1}{4}$% = 7.25%

Formula	*Cross multiply*	*Divide by 100*
$\frac{7.25}{100}$ ⤫ $\frac{?}{6}$	7.25 × 6 = 43.5	43.5 ÷ 100 = 43.5 = 0.435

Answer: 0.435 is $7\frac{1}{4}$% of 6.

CHECKPOINT 3-1

YOUR GOAL:
Get 5 or more answers correct.

Solve the following problems. Round to the nearest hundredth. Work each problem on a separate piece of paper. Write your answers in the spaces provided. The first one has been done as an example.

● What is 25% of 8.41?   __2.10__

$\frac{25}{100}$ ⤫ $\frac{?}{8.41}$ 25 × 8.41 = 210.25   210.25 ÷ 100 = 2.10

1. What is 15% of 28?   _____

2. What is 5% of $15.00?   _____

3. What is $7\frac{1}{2}$% of 12?   _____

4. 21% of 32 is?   _____

5. $6\frac{1}{2}$% of 38 is?   _____

6. 10% of 52 is?   _____

 Check your work on page 43. Record your score on page 47.

IT'S A FACT—Catalog Sales

Catalog sales are booming. Surveys show over half the people in the United States purchased something by mail last year. Some part of catalog sales may have occupational opportunities.

FINDING WHAT PERCENT ONE NUMBER IS OF ANOTHER

The same formula is used when you want to find the percent of a number. Cross multiply then divide, as shown in Illustration 3-5.

Illustration 3-5

What percent of 150 is 45?

Formula	Step 1:	Step 2:
	Cross multiply	Divide the result of cross multiplying by the remaining number
$\dfrac{\%}{100} = \dfrac{Part}{Total}$	$\dfrac{?}{100} = \dfrac{45}{150}$	
	$100 \times 45 = 4{,}500$	$4{,}500 \div 150 = 30$

Answer: 30% of 150 is 45.

The quick way to multiply by 100 is to move the decimal point two places to the right. Multiplying by 100 makes the number larger. Compare the two ways of multiplying by 100 in Illustration 3-6.

Illustration 3-6

$$
\begin{array}{r}
253 \\
\times\ 100 \\
\hline
000 \\
000 \\
253 \\
\hline
25{,}300
\end{array}
$$

$253.00 \times 100 = 253.00 = 25{,}300$

Add 2 zeros   Move decimal

CHECKPOINT 3-2

YOUR GOAL:
Get 5 or more answers correct.

Solve the following problems. Use a separate piece of paper to work each problem. Write your answers in the spaces provided. The first one has been done as an example.

● What percent of 250 is 35?   __14%__

$\dfrac{\%}{100} = \dfrac{35}{250}$ $35 \times 100 = 3{,}500$ $3{,}500 \div 250 = 14\%$

1. What percent of 80 is 20?   _____

2. What percent of 400 is 60?   _____

3. What percent of 24 is 12?   _____

4. 15 is what percent of 60?   _____

5. 25 is what percent of 125?   _____

6. 12 is what percent of 48?   _____

 Check your work on page 43. Record your score on page 47.

◀ **IT'S A FACT—Child Care Costs**

Annual child care costs range from about $3,000 to $5,500 depending on the area in which you live and the age of the child. This amount of money represents between 6 and 12 percent of the annual income of a family earning $40,000 to $50,000.

FINDING A NUMBER WHEN THE PERCENT OF THE NUMBER IS KNOWN

The same formula is used to *find the number* when the percent is known. Cross multiply then divide, as shown in Illustration 3-7.

Illustration 3-7

12 is 48% of what number?

$\dfrac{48}{100}$ ⤴ $\dfrac{12}{?}$ $100 \times 12 = 1200$ $1200 \div 48 = 25$

Answer: 12 is 48% of 25.

✔ *CHECKPOINT 3-3*

YOUR GOAL:
Get 5 or more answers correct.

Solve the following problems. Use a separate piece of paper to work each problem. Write your answers in the spaces provided. The first one has been done as an example.

● 52% of what number is 156?   __300__

$\dfrac{52}{100}$ ⤴ $\dfrac{156}{?}$ $100 \times 156 = 15,600$ $15,600 \div 52 = 300$

1. 10 is 25% of what number?   _____

2. 3 is 50% of what number?   _____

3. 75% of what number is 15?   _____

4. 60 is 25% of what number?   _____

5. 40% of what number is 16?   _____

6. 18 is 6% of what number?   _____

☞ *Check your work on page 43. Record your score on page 47.*

FIGURING PERCENT WORD PROBLEMS

It is sometimes a challenge to decide how to set up a percent word problem. On the previous pages you learned that the same formula could be used for all three types of percent figuring. Where to put the number in the formula is the challenge. Illustration 3-8 shows one way that may help you.

Illustration 3-8

Put % over 100   The number after the word *of* is the denominator   The number after the word *is* becomes the numerator

$$50\% \text{ of } 6 \text{ is } 3$$

$$\frac{50}{100} = \frac{\text{is } 3}{\text{of } 6}$$

In Illustration 3-9, a word problem is studied to determine where to put the numbers in the formula.

Illustration 3-9

Jim ate 10 of the cookies he baked. He ate 25% of the cookies. How many cookies did Jim bake?

% always goes over 100

$$\frac{25}{100} = \frac{10}{?}$$ Part of cookies Jim ate
Total cookies Jim baked

There isn't a number after the word *of,* so that is what we are trying to find. 10 cookies *is* 25%.

✔ ## CHECKPOINT 3-4

YOUR GOAL:
Get 5 or more answers correct.

Solve the following percent problems on a separate piece of paper. Write your answers in the spaces provided. The first one has been done as an example.

● Kim and Yoko got a box of candy as a present. Kim ate 18 out of 24 candies. What percent of the candies did Kim eat?

 $\frac{\%}{100} = \frac{18}{24}$ $100 \times 18 = 1,800$ $1,800 \div 24 = 75\%$ **75%**

1. The women's club sold cookbooks for a fund-raiser. They had 200 books and sold 80% the first month. How many cookbooks did they sell the first month?

   _____

2. A fabric store is having a 40% off sale on cotton fabrics. The fabric was reduced by $3.50. What was the original price?

   _____

3. A manager was checking his emplyoyees' percentage of absences for the last 90 days. Lee was absent 9 days. What percent was she absent?

   _____

4. A farmer has 350 acres planted in wheat. He was able to harvest 30% of his 350 acres before it rained. How many acres did he harvest?

   _____

5. Thirty-six percent of the employees at the factory are women. If 450 women work at the factory, what is the total number of employees?

   _____

6. The survey showed that 35 out of the 140 people questioned favored the redevelopment project. What percent favored the project?

   _____

☞ *Check your work on page 43. Record your score on page 47.*

WHAT YOU HAVE LEARNED

Now that you have finished this unit you will be able to find the percent of a number, find what percent one number is of another, and find the number when the percent is known. Several methods for calculating percent word problems were presented. You also learned an easy way to multiply and divide by 100.

ACTIVITY 3-1   **YOUR GOAL:** Get 5 or more answers correct.

Solve the following problems. Round to the nearest hundredth. Work each problem on a separate piece of paper. Write your answers in the spaces provided. The first one has been done as an example.

- What is 4% of $600?   __**24**__

  $\frac{4}{100} = \frac{?}{600}$ $4 \times 600 = 2,400$ $2,400 \div 100 = 24$

1. What is 12% of 1,125?   _____

2. What is 25% of 16?   _____

3. What is 9% of 81?   _____

4. 5% of 640 is   _____

5. 50% of 124 is   _____

6. 20% of 160 is   _____

☞ *Check your work on page 43. Record your score on page 47.*

ACTIVITY 3-2   **YOUR GOAL:** Get 5 or more answers correct.

Solve the following problems. Use a separate piece of paper to work each problem. Write your answers in the spaces provided. The first one has been done as an example.

- What percent of 500 is 125?   __**25**__

  $\frac{\%}{100} = \frac{125}{500}$ $100 \times 125 = 12,500$ $12,500 \div 500 = 25$

1. What percent of 30 is 3?   _____

2. What percent of 600 is 180?   _____

3. What percent of 200 is 25?   _____

4. 5 is what percent of 25?   _____

5. 18 is what percent of 72?   _____

6. 26 is what percent of 104?   _____

☞ *Check your work on page 43. Record your score on page 47.*

ACTIVITY 3-3   YOUR GOAL: Get 5 or more answers correct.

Solve the following problems. Use a separate piece of paper to work each problem. Write your answers in the spaces provided. The first one has been done as an example.

● 12 is 25% of what number?   **48**

$\frac{25}{100} \rightarrow \frac{12}{?}$ 100 × 12 = 1,200   1,200 ÷ 25 = 48

1. 14 is 50% of what number?   _____

2. 450 is 75% of what number?   _____

3. 24 is 4% of what number?   _____

4. 30% of what number is 30?   _____

5. 9% of what number is 18?   _____

6. 15% of what number is 13.5?   _____

☞ **Check your work on page 43. Record your score on page 47.**

ACTIVITY 3-4   YOUR GOAL: Get 5 or more answers correct.

Solve the following percent problems on a separate piece of paper. Write your answers in the spaces provided. The first one has been done as an example.

● A business had a 20% gain in sales. The gain was $600. What was the store's total sales?

$\frac{20}{100} \rightarrow \frac{600}{?}$ 100 × 600 = 60,000   60,000 ÷ 20 = $3,000   **$3,000**

1. Thirty-three percent of the people in Longtown said they would vote for the school bond. There are 22,000 people in Longtown. How many said they would vote for the school bond?

_____

2. The credit union charged 12% interest on a car loan. The interest was $1,800. What was the amount of the car loan?

_____

3. Martin paid $4,200 in income tax. His total wages were $28,000. What percent tax did he pay?

_____

4. A store was having a 40% off sale on garments containing the color blue. What is the discount on a $28.00 dress?

_____

5. Thirty percent of the store's sales is $2,700. What is the total sales?

_____

6. Six out of 12 members of Mr. Parrott's basketball team were on time for practice. What percent were on time?

_____

☞ *Check your work on page 43. Record your score on page 47.*

CHECKING WHAT YOU LEARNED

Now you can see how much you have learned about percents. These 25 questions cover the main topics you studied in this book. There is no time limit, so take your time.

Follow the directions carefully. Write your answers in the spaces provided. When you finish, check your answers. Give yourself 4 points for each correct answer. Record your score in your Personal Progress Record. The analysis chart will help you tell where you may need additional study.

DIRECTIONS: Change the following amounts to decimals, fractions, and percents.

	Decimal	**Fraction**	**Percent**	● ___0.05___
Five hundredths =	● 0.05	**(1)**	**(2)**	1. _____
				2. _____
Forty hundredths =	**(3)**	**(4)**	**(5)**	3. _____
				4. _____
Three tenths =	**(6)**	**(7)**	**(8)**	5. _____
				6. _____
Four thousandths	**(9)**	**(10)**	**(11)**	7. _____
				8. _____

Change to percents.

9. _____

(12) $\frac{4}{5}$ **(13)** $\frac{2}{5}$ **(14)** $\frac{1}{4}$ **(15)** $\frac{1}{2}$

10. _____

11. _____

12. _____

13. _____

14. _____

15. _____

Work the following problems.

(16) 25% of 200   **(17)** 5% of 40   **(18)** 15% of 60

(19) What percent of 60 is 3?

(20) What percent of 120 is 30?

(21) 10 is 25% of what number?

(22) 80 is 40% of what number?

(23) How much would a 15% tip be on a $12.00 dinner?

(24) Maury got 20% off on a pair of shoes that regularly sells for $40.00. How much did he pay for the shoes?

(25) Mrs. Taylor figured that 80% of the 30 people on her committee could attend the meeting. How many people could attend the meeting?

☞ *Check your work on page 44. Record your score on page 48.*

GLOSSARY

E

Equivalent Values that are equal.

P

Percent A part of something that has been divided into 100 parts.

INDEX

1. 7/100
2. 7%
3. 0.30
4. 30/100
5. 30%
6. 0.1
7. 1/10
8. 10%
9. 0.009
10. 9/1000
11. 0.9%
12. 37.5%
13. 60%
14. 50%
15. 62.5%
16. 45
17. 2.40
18. 8.10
19. 25%
20. 20%
21. 60
22. 140
23. $1.20
24. $787.50
25. 20 games

UNIT 1

CHECKPOINT 1-1 page 2

1. A. $\frac{15}{100} = 15\%$
   B. $\frac{3}{100} = 3\%$
   C. $\frac{75}{100} = 75\%$
2. A. 40%
   B. 35%
   C. 47%
3. A. 20%
   B. 10%

CHECKPOINT 1-2, page 4

1. 20%
2. 150%
3. 75%
4. 3%
5. 200%
6. 84%
7. 30%

CHECKPOINT 1-3, page 5

1. 12%
2. 235%
3. 350%
4. 38%
5. 7%
6. 100%
7. 90%

CHECKPOINT 1-4, page 6

1. 0.21
2. 2.50
3. 0.03
4. 0.40
5. 1.15

CHECKPOINT 1-5, page 7

1. five tenths $\frac{5}{10}$
2. thirty thousandths $\frac{30}{1000}$
3. twenty-five hundredths $\frac{25}{100}$
4. ninety-nine hundredths $\frac{99}{100}$
5. seven tenths $\frac{7}{10}$
6. one hundred twenty-eight thousandths $\frac{128}{1000}$
7. nine hundredths $\frac{9}{100}$
8. one tenth $\frac{1}{10}$

CHECKPOINT 1-6, page 8

1. A. $\frac{20}{100} = 20\%$
   B. $\frac{40}{100} = 40\%$
   C. $\frac{16}{100} = 16\%$
2. A. $\frac{50}{100} = \frac{1}{2}$
   B. $\frac{60}{100} = \frac{3}{5}$
   C. $\frac{20}{100} = \frac{1}{5}$

CHECKPOINT 1-7, page 9

1. 40%
2. 42.9%
3. 83.3%

CHECKPOINT 1-8, page 10

1. 0.12125
2. 0.064
3. 0.065
4. 0.7575
5. 0.20375

ACTIVITY 1-1, page 11

1. A. $\frac{30}{100} = 30\%$
   B. $\frac{3}{100} = 3\%$
   C. $\frac{85}{100} = 85\%$
2. A. 50%
   B. 8%
   C. 17%
3. 8%

ACTIVITY 1-2, page 12

1. 65%
2. 30%
3. 3%

ACTIVITY 1-3, page 12

51. 9%
2. 200%
3. 370%

ACTIVITY 1-4, page 12

1. 1.28
2. 0.02
3. 0.27

ACTIVITY 1-5, page 13

1. Seven tenths $\frac{7}{10}$
2. Twenty-seven hundredths $\frac{27}{100}$
3. Forty hundredths $\frac{40}{100}$
4. Five hundredths $\frac{5}{100}$

ACTIVITY 1-6, page 13

1. A. $\frac{60}{100}$ = 60%
   B. $\frac{20}{100}$ = 20%
   C. $\frac{35}{100}$ = 35%
2. A. $\frac{80}{100}$ = $\frac{4}{5}$
   B. $\frac{5}{100}$ = $\frac{1}{20}$
   C. $\frac{13}{100}$

ACTIVITY 1-7, page 14

1. 38%
2. 17%
3. 33%

ACTIVITY 1-8, page 14

1. 0.035
2. 0.096
3. 0.4075 = 0.408
4. 0.2129 = 0.213
5. 2.0133 = 2.013

UNIT 2

CHECKPOINT 2-1, page 16

1. $0.09
2. 7.0
3. $75.00
4. 30.0
5. $28.00
6. 11.25

CHECKPOINT 2-2, page 17

1. 42.9
2. 370
3. 0.02
4. 1.1
5. 4.5
6. 5.8

CHECKPOINT 2-3, page 18

1. 40
2. 50
3. 202
4. 190
5. 17.5
6. 10

CHECKPOINT 2-4, page 19

1. 40
2. 100
3. 200
4. 400
5. 30
6. 75
7. 150
8. 3

CHECKPOINT 2-5, page 20

1. $94.79
2. $3.45
3. $5,600
4. $480.00
5. $1,582.50
6. $1.80

ACTIVITY 2-1, page 22

1. 16
2. 5.6
3. 40
4. 52.5
5. 3.5
6. 400

ACTIVITY 2-2, page 22

1. 78.75
2. 0.25
3. 2.05
4. 19.4
5. 10.25
6. 7.29

Answers

ACTIVITY 2-3, page 23

1. 5
2. 12.5
3. 150
4. 250
5. 20
6. 45

ACTIVITY 2-4, page 23

1. 3
2. 200
3. 13
4. 380
5. 50
6. 125
7. 250
8. 5

ACTIVITY 2-5, page 24

1. $1.47
2. $14.38
3. $1,500
4. $10.50
5. $71.30
6. $37.63

UNIT 3

CHECKPOINT 3-1, page 27

1. 4.20
2. $0.75
3. 0.90
4. 6.72
5. 2.47
6. 5.20

CHECKPOINT 3-2, page 28

1. 25%
2. 15%
3. 50%
4. 25%
5. 20%
6. 25%

CHECKPOINT 3-3, page 29

1. 40
2. 6
3. 20
4. 240
5. 40
6. 300

CHECKPOINT 3-4, page 30

1. 160
2. $8.75
3. 10%
4. 105
5. 1,250
6. 25%

ACTIVITY 3-1, page 32

1. 135
2. 4
3. 7.29
4. 32
5. 62
6. 32

ACTIVITY 3-2, page 32

1. 10%
2. 30%
3. 12.5%
4. 20%
5. 25%
6. 25%

ACTIVITY 3-3, page 33

1. 28
2. 600
3. 600
4. 100
5. 200
6. 90

ACTIVITY 3-4, page 33

1. 7,260
2. $15,000
3. 15%
4. $11.20
5. $9,000
6. 50%

✔ **CHECKING WHAT YOU LEARNED**

1. 5/100
2. 5%
3. 0.40
4. 40/100
5. 40%
6. 0.3
7. 3/10
8. 30%
9. 0.004
10. 4/1000
11. .4%
12. 80%
13. 40%
14. 25%
15. 50%
16. 50
17. 2
18. 9
19. 5%
20. 25%
21. 40
22. 200
23. $1.80
24. $32.00
25. 24 people

Name: _____

✔ CHECKING WHAT YOU KNOW—ANALYSIS CHART

Use the chart below to determine the areas in which you may need additional study. In the space provided, write the total number of points you got for each content area. Each correct answer is worth 4 points. Then add the total number of points to find your final score. Circle the items you answered incorrectly. Use the page references for those items to locate the topics. As you begin your study, pay close attention to those areas where you missed half or more of the questions.

Content Area	Item Number	Study Pages	Total Points	Number Right
Equivalents: decimals, fractions, and percents	1, 2, 3, 4, 5, 6, 7, 8, 9, 10, 11	5–10	44	
Changing fractions to percents	12, 13, 14, 15	5–10	16	
Finding the percent of a number	16, 17, 18	15–17	12	
Finding what percent one number is of another	19, 20	27–28	8	
Finding a number when the percent of the number is known	21, 22	29	8	
Percent word problems	23, 24, 25	30	12	

Date _____   Total Points: 100   Your Score: _____

Unit 1 Understanding Percents

Exercise	Score
Checkpoint 1-1	_____
Checkpoint 1-2	_____
Checkpoint 1-3	_____
Checkpoint 1-4	_____
Checkpoint 1-5	_____
Checkpoint 1-6	_____
Checkpoint 1-7	_____
Checkpoint 1-8	_____
Activity 1-1	_____
Activity 1-2	_____
Activity 1-3	_____
Activity 1-4	_____
Activity 1-5	_____
Activity 1-6	_____
Activity 1-7	_____
Activity 1-8	_____
Total, Unit 1	_____

HOW ARE YOU DOING?

73 or 95	Excellent
68 to 72	Good
62 to 67	Fair
Less than 62	See instructor

Unit 2 Finding the Percent of a Number

Exercise	Score
Checkpoint 2-1	_____
Checkpoint 2-2	_____
Checkpoint 2-3	_____
Checkpoint 2-4	_____
Checkpoint 2-5	_____
Activity 2-1	_____
Activity 2-2	_____
Activity 2-3	_____
Activity 2-4	_____
Activity 2-5	_____
Total, Unit 2	_____

HOW ARE YOU DOING?

54 or better	Excellent
48 to 53	Good
42 to 47	Fair
Less than 42	See instructor

Unit 3 Finding Other Kinds of Percents

Exercise	Score
Checkpoint 3-1	_____
Checkpoint 3-2	_____
Checkpoint 3-3	_____
Checkpoint 3-4	_____
Activity 3-1	_____
Activity 3-2	_____
Activity 3-3	_____
Activity 3-4	_____
Total, Unit 3	_____

HOW ARE YOU DOING?

40 or better	Excellent
35 to 39	Good
62 to 67	Fair
Less than 30	See instructor

✔ CHECKING WHAT YOU LEARNED—ANALYSIS CHART

Use the chart below to determine the areas in which you need additional study. In the space provided, write the total number of points you got for each content area. Each correct answer is worth 4 points. Then add the total number of points to find your final score. Circle the items you answered incorrectly. Use the page references for those items to locate the topics. Go back to those pages and review the items missed.

Content Area	Item Number	Study Pages	Total Points	Number Right
Equivalents: decimals, fractions, and percents	1, 2, 3, 4, 5, 6, 7, 8, 9, 10, 11	5–10	44	
Changing fractions to percents	12, 13, 14, 15	5–10	16	
Finding the percent of a number	16, 17, 18	15–17	12	
Finding what percent one number is of another	19, 20	27–28	8	
Finding a number when the percent of the number is known	21, 22	29	8	
Percent word problems	23, 24, 25	30	12	

Date _____

Total Points: 100   Your Score: _____